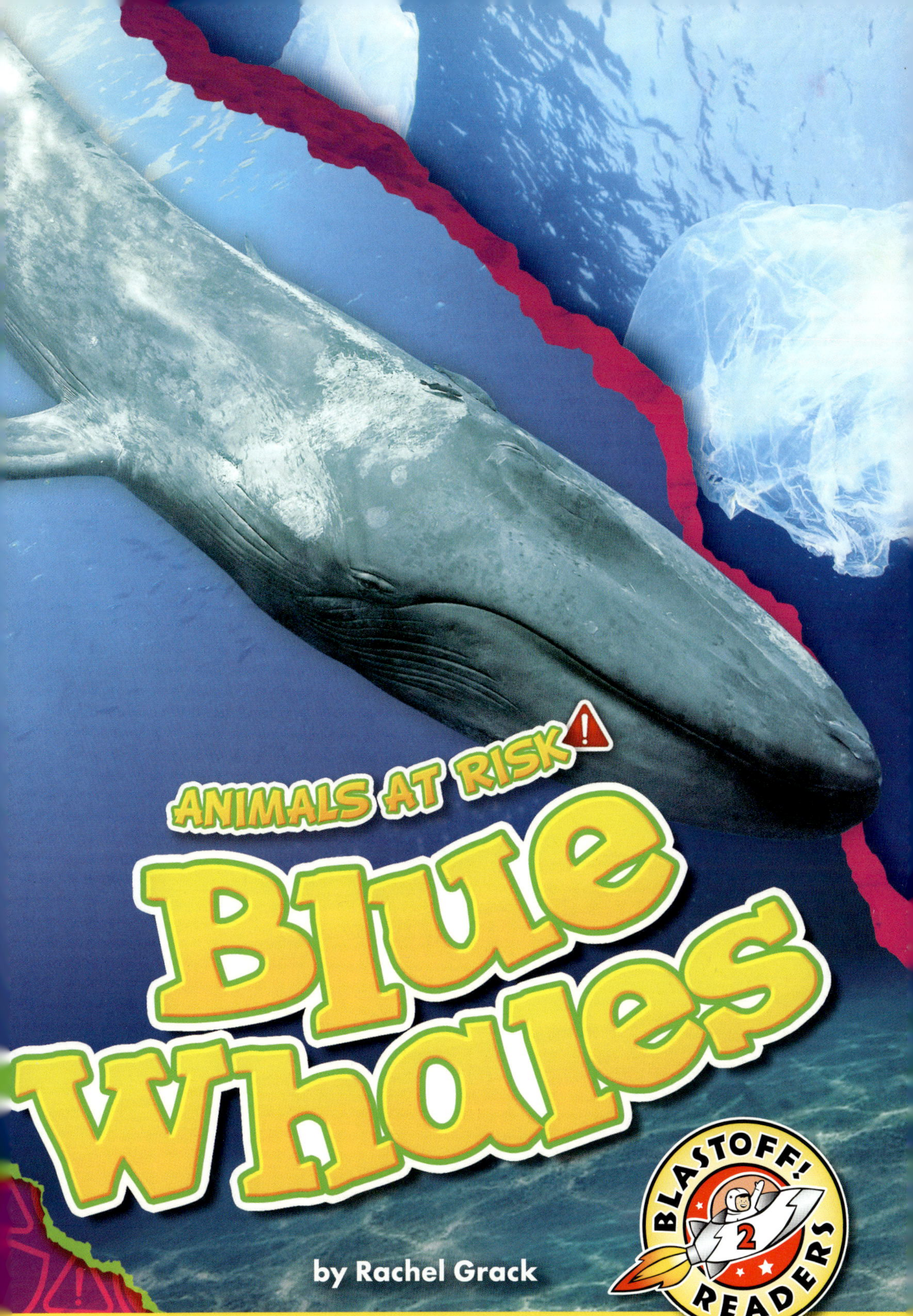

ANIMALS AT RISK

Blue Whales

by Rachel Grack

BLASTOFF! READERS 2

BELLWETHER MEDIA • MINNEAPOLIS, MN

Blastoff! Readers are carefully developed by literacy experts to build reading stamina and move students toward fluency by combining standards-based content with developmentally appropriate text.

Level 1 provides the most support through repetition of high-frequency words, light text, predictable sentence patterns, and strong visual support.

Level 2 offers early readers a bit more challenge through varied sentences, increased text load, and text-supportive special features.

Level 3 advances early-fluent readers toward fluency through increased text load, less reliance on photos, advancing concepts, longer sentences, and more complex special features.

★ Blastoff! Universe

This edition first published in 2023 by Bellwether Media, Inc.

Library of Congress Cataloging-in-Publication Data

Names: Koestler-Grack, Rachel A., 1973- author.
Title: Blue whales / Rachel Grack.
Description: Minneapolis, MN : Bellwether Media, 2023. | Series: Blastoff! readers: animals at risk | Includes bibliographical references and index. | Audience: Ages 5-8 | Audience: Grades 2-3 | Summary: "Relevant images match informative text in this introduction to blue whales. Intended for students in kindergarten through third grade"-- Provided by publisher.
Identifiers: LCCN 2022000415 (print) | LCCN 2022000416 (ebook) | ISBN 9781644877104 (library binding) | ISBN 9781648347566 (ebook)
Subjects: LCSH: Blue whale--Juvenile literature. | Blue whale--Conservation--Juvenile literature.
Classification: LCC QL737.C424 K64 2023 (print) | LCC QL737.C424 (ebook) | DDC 599.5/248--dc23/eng/20220113
LC record available at https://lccn.loc.gov/2022000415
LC ebook record available at https://lccn.loc.gov/2022000416

Editor: Kieran Downs Designer: Brittany McIntosh

Printed in the United States of America, North Mankato, MN.

Table of Contents

Huge Swimmers

Blue whales are the largest animals on Earth! Their blue-gray skin looks blue underwater.

Five **subspecies** swim in oceans around the world.

Years ago, people hunted blue whales for **whale oil**. They became **endangered**.

People continue to put these **mammals** in danger today.

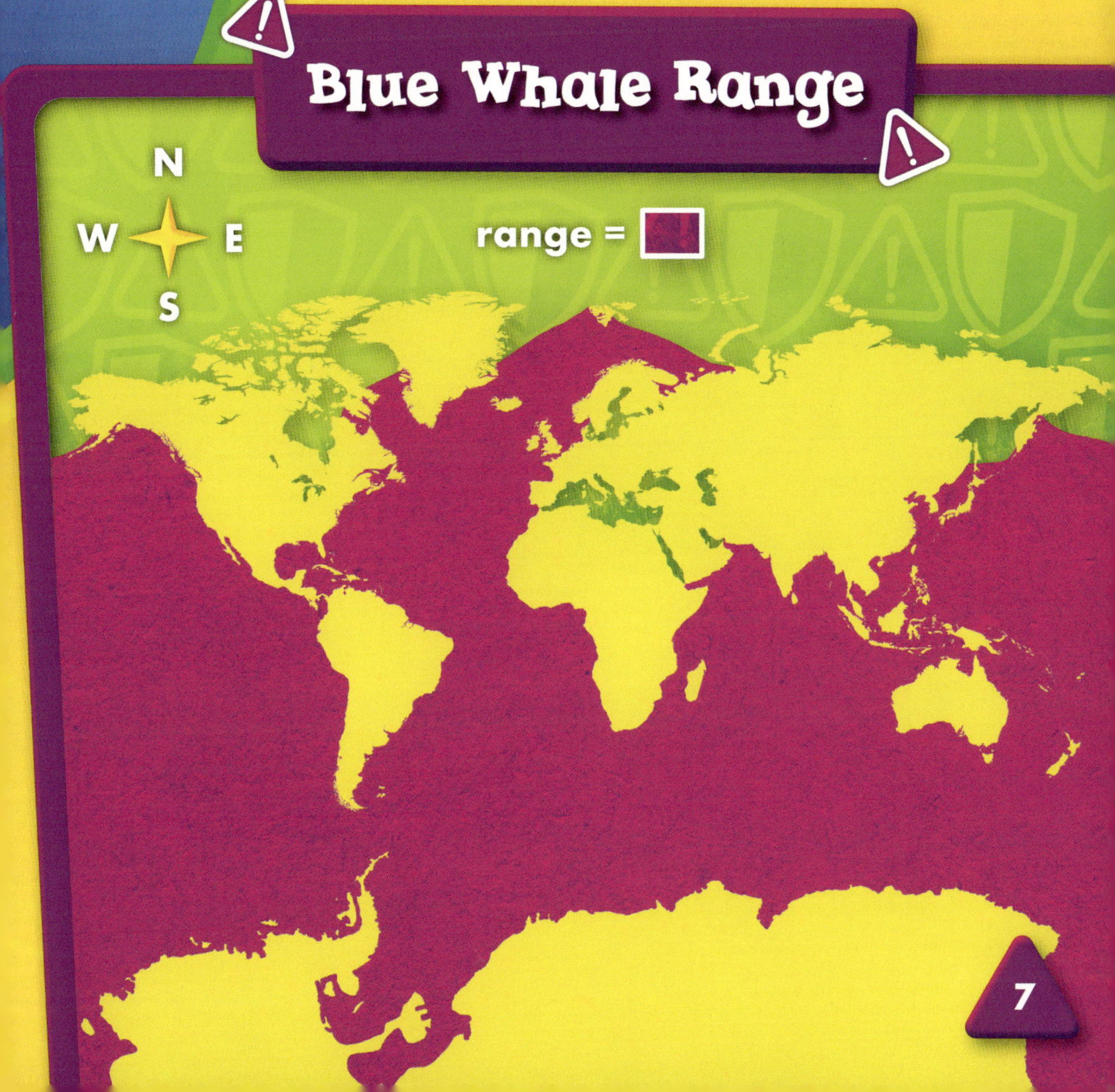

In Danger!

Shipping lanes often run through blue whale **habitats**. Ships may hit the whales.

Blue whales also get trapped in fishing gear.

Threats

1 humans need to move goods

2 ships travel across oceans

3 ships hit blue whales

Oceans are filled with **pollution**. Blue whales sometimes think plastic is food.

Climate change is warming waters. This harms **krill**. Blue whales have less to eat.

Blue Whale Stats

Least Concern	Near Threatened	Vulnerable	Endangered	Critically Endangered	Extinct in the Wild	Extinct

conservation status: endangered

life span: around 80 to 90 years

Save the Blue Whales!

Without blue whales,
all sea life would suffer.

The whales keep the **food chain** healthy. Their poop puts **nutrients** into the ocean **ecosystem**.

The World with Blue Whales

Today, it is against the law to hunt blue whales.

People study whale **migrations** and habitats. Large ships avoid their homes.

Sailors watch for blue whales. They use tools to spot them underwater.

Ships slow down and carefully steer around the whales.

People study climate change. They learn how it affects blue whales.

blue whale
caught in fishing gear

People also help blue whales in trouble. They free whales caught in fishing gear.

People help blue whales by throwing away trash correctly. Using less **energy** can slow climate change.

Together, everyone can save these huge animals!

Glossary

climate change–a human-caused change in Earth's weather due to warming temperatures

ecosystem–a community of plants and animals living in a certain place

endangered–in danger of dying out

energy–usable power that comes from things including heat, electricity, and sound

food chain–the order of animals and plants in which each feeds on the one below it in the chain

habitats–places and natural surroundings in which plants or animals live

krill–small shrimp-like sea animals; krill are the main food of blue whales.

mammals–warm-blooded animals that have backbones and feed their young milk

migrations–trips from one place to another, often with the seasons

nutrients–things needed by people, animals, and plants to stay strong and healthy

pollution–substances that make nature dirty; pollution usually comes from humans.

subspecies–different kinds of one type of animal

whale oil–oil from whale fat once used in oil lamps and soap

To Learn More

AT THE LIBRARY

Chanez, Katie. *Blue Whales.* Mendota Heights, Minn.: Apex, 2021.

Kenney, Karen Latchana. *Oceans.* Minneapolis, Minn.: Bellwether Media, 2022.

Rossiter, Brienna. *Saving Earth's Animals.* Lake Elmo, Minn.: Focus Readers, 2022.

ON THE WEB

FACTSURFER

Factsurfer.com gives you a safe, fun way to find more information.

1. Go to www.factsurfer.com.

2. Enter "blue whales" into the search box and click 🔍.

3. Select your book cover to see a list of related content.

Index

The images in this book are reproduced through the courtesy of: WaterFrame/ Alamy Stock Photo, front cover, pp. 3, 6, 14; Rich Carey, front cover (back top), pp. 3, 23; A Cotton Photo, front cover (back bottom); Tony Wu/ Nature Picture Library, pp. 4, 8; Nature Picture Library/ Alamy Stock Photo, pp. 5, 12, 19, 20-21, 23; gyn9037, p. 9 (top left); Aerial-motion, p. 9 (top right); Flip Nicklin/ Minden Pictures, pp. 9 (bottom), 18; RLS Photo, p. 10; DOC WHITE/ Nature Picture Library, pp. 10-11; Chase Dekker, p. 13 (top left); BIOSPHOTO/ Alamy Stock Photo, p. 13 (top right); W_underwater, p. 13 (bottom); Anthony Pierce/ Alamy Stock Photo, p. 15; Andrew Sutton, p. 16; sbedaux, p. 17; fotosparrow, p. 20.